Grades 5–6

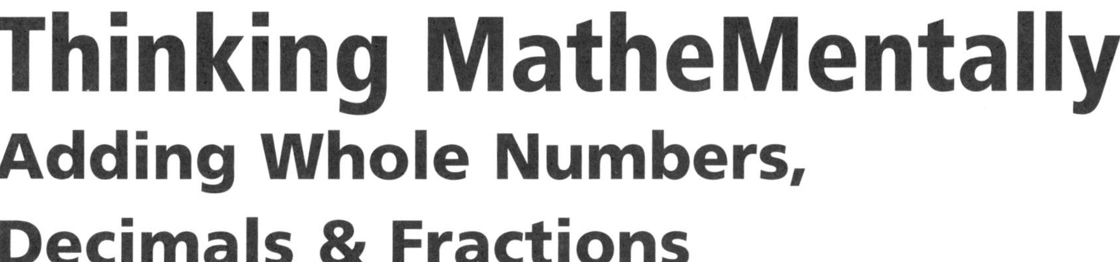

Thinking MatheMentally

Adding Whole Numbers, Decimals & Fractions

James Burnett

Series Consultants:
Terry Rose
Kaycie Hersey
Beverley Neitzel

Wright Group
McGraw-Hill

Thinking MatheMentally
Whole Numbers, Decimals and Fractions

Text: James Burnett
Project editor: Beth Lewis
Illustrations: Aden Jacobi
Page layouts: Robin Hutchens

Published by
PRIME EDUCATION
A division of JAMCAL Corporation Pty Ltd.
P O Box 74
Narangba QLD 4504 Australia
www.primeeducation.com

Exclusively distributed in the USA by
Creative Publications, an imprint of Wright Group/McGraw-Hill
19201 120th Avenue NE, Suite 100
Bothell, W 98011
www.WrightGroup.com

Printed in Australia

2006 2005 2004 2003 2002
10 9 8 7 6 5 4 3 2 1

ISBN: 1 876842 90 3

Contents

Introduction

For many years, written paper-and-pencil computation has been the focus of school mathematics programs. As a result, students often apply a written algorithm when it is far easier, quicker and more appropriate to perform the computation mentally. Students gain many benefits by becoming skilled at mental computation. Not only do they develop a better understanding of operations, place value and number properties, but they gain a better sense of numbers, greater confidence in their abilities and improved attitudes towards mathematics. Students should be given regular opportunities to develop their mental computation skills and encouraged to apply them whenever they are appropriate.

What is mental computation?

Mental computation refers to the calculation of an exact answer 'in one's head'. It therefore makes no use of external tools such as paper and pencils or calculators. The method used to obtain the answer is generally invented but it can also be borrowed from a standard written algorithm. Mental computation is more than daily 'warm ups'. Activities such as this, which encourage speedy responses, constitute testing rather than teaching and are usually labelled 'mental arithmetic'. Mental computation, on the other hand, focuses on the thinking processes adopted in the strategy rather than on the product as in mental arithmetic.

How do we develop mental computation?

Although the benefits of a curriculum that encourages mental computation are widely accepted, the research literature does not reveal a preferred methodology for developing mental computation. However, there is evidence that certain teaching practices encourage students to develop good mental computation skills. First, research shows that understanding is enhanced when opportunities are provided for students to share and justify their answers and thinking strategies. Therefore, an effective approach focuses on discussion and on the sharing of devised mental strategies with the teacher and peers.

Second, McIntosh (et al, 1995) and Reys (et al, 1995) found that the mode of presentation affected students' ability to compute mentally. In their studies, there were times when students performed better with tasks presented orally and on other occasions they performed better with tasks presented visually. An effective approach provides a variety of representational models so that students are given every opportunity for success and to develop the ability to switch from one model to another.

Finally, consideration must be given to a context when presenting tasks. Students' natural curiosity can be aroused by involving them in activities that require mental computation. A real-world situation involving money often triggers a different approach than one that is totally devoid of a context. For example, it seems natural to think of a CD as costing $20.00 less 5 cents, and 3 CDs as $60.00 less 15 cents. An effective approach presents carefully chosen number sentences in a context that stimulates discussion.

About this series

Thinking MatheMentally is designed to help students develop a range of mental computation strategies for whole numbers, decimals and fractions. The activities focus on mental strategies that are often used in everyday life. These addition, subtraction, multiplication and division strategies have been carefully sequenced throughout the series to enable students to build their confidence in mental computation.

About this book

This book is divided into five units. Each unit introduces and develops a different strategy. The units are carefully sequenced so the work of one unit builds on or uses ideas from a previous unit.

Each unit has four focus activities and one review activity. Each is expected to take about twenty minutes to complete. A teacher may choose to complete one unit in one week. In doing so, this book provides five weeks of strategy building for addition.

This book is consistent with the effective teaching approach described on page 4. Two pages are dedicated to each activity. The left-hand page provides the teachers' notes and the right-hand page is designed to be reproduced as an overhead transparency (OHT).

The OHT is the focus of the activity. It provides a problem set in a context to stimulate discussion. At this stage, the students are encouraged to share their strategies to solve the problem. It should be noted that no one strategy is the right strategy. All suggestions should be valued. A suitable mental strategy for solving the problem is then explained on the OHT. Further examples are also provided to discuss as a whole class. The students conclude the activity by completing written examples in their individual 'Mastery Booklet'. This booklet is made from a photocopy of a blackline master (see pages 56–60). Give the students a copy of the blackline master at the end of the first activity in each unit. Show them how to fold the page in half and in half again to make the booklet.

A 25-item written test is provided as Blackline Master 7 on page 62. It provides examples that require different mental strategies. This test could be administered as an oral or written pretest or posttest. Remember there is no one correct strategy for any given example. The students should be encouraged to use any strategy that suits their way of thinking.

References

MacIntosh, A. J., Bana, J. & Farrell, B. (1995). *Mental computation in school mathematics: Preference, attitude and performance of students in years 3, 5, 7 and 9.* Perth, WA: MASTEC.

Reys, R. E., Reys, B. J., Nohda, N. & Emori, H. (1995). Mental computation performance and strategy use of Japanese students in grades 2, 4, 6 and 8. *Journal for Research in Mathematics Education,* 26(4), 304–326.

Adding two- and three-digit numbers – with bridging

Introduction

In this activity, the students use a 'left-to-right' strategy to add number sentences such as 325 + 47 that require bridging or 'jumping' across a ten or hundred. For example 325 + 47 is calculated by adding (320 + 40) + (5 + 7). The students are also encouraged to discuss other strategies for calculating the answer mentally.

Procedure

1. Show Overhead Transparency 1.1 (OHT 1.1) and invite a volunteer to read the problem at the top. Lead a discussion about possible ways to calculate the answer mentally. Ask, *What strategy could we use to calculate the answer?* The discussion will vary but the students may suggest breaking the second addend into tens and ones, for example 325 + 40 + 7. Another strategy might be to change 47 to 45 and adjust the answer (325 + 47 is the same as 325 + 45 + 2). Write the students' suggestions on the board.

2. Read the strategy described in the thought balloon. Call upon a volunteer to explain the strategy in his/her own words. Afterwards, have the students compare the strategy with those listed on the board. Ask, *Which strategy seems more appropriate for this example? Are any strategies difficult to understand? Why?* Remember that no one strategy is the correct strategy. The students should be encouraged to choose and use the method that best suits their way of thinking.

3. Discuss the number sentences at the bottom of the OHT. For each example, ask a volunteer to verbalize the strategy as he/she calculates the answer. Examples 7, 8 and 9 may be more challenging, as these require the students to bridge over a hundred.

4. Give the students a copy of Blackline Master 1 (page 56). Show them how to fold the page in half and in half again to make a small booklet. Allow time for the students to finish the ten examples for Activity 1.1.

Adding left to right

The circus presold 325 tickets for opening night. Another 47 tickets were sold at the door. How many tickets were sold altogether?

Mmm ... 325 + 47

320 + 40 = 360 *and*
5 + 7 = 12 *so*
360 + 12 = 372

Try adding from the left to calculate each of these.

1. **425 + 36**	4. **34 + 247**	7. **526 + 81**
2. **255 + 38**	5. **264 + 28**	8. **622 + 87**
3. **155 + 27**	6. **46 + 337**	9. **67 + 351**

Adding three-digit numbers – with bridging

Introduction

A 'left-to-right' strategy was investigated in the addition book in each of the previous levels of *Thinking MatheMentally.* In this activity, the students use this strategy to add 3-digit numbers that require bridging over a ten or hundred. For example, 425 + 268 is calculated by adding 400 + 200) + (20 + 60) + (5 + 8).

Procedure

1. Show and read the problem at the top of OHT 1.2. Challenge the students to calculate the answer mentally.

2. Show and read the speech balloon to the students. Survey the class to determine the number of students who used this 'left-to-right' strategy. It is possible that some students may use this strategy a little differently. They may choose to add all the tens together — 425 + 268 is the same as 42 tens + 26 tens + 13 ones. Flexible thinking such as this should be praised and encouraged. Invite those who used a different strategy to explain their methods to the class. For example, some students may follow a traditional paper-and-pencil algorithm and add from the right first. Others may break the second **addend** into hundreds, tens and ones to add 425 + 200 + 60 + 8. Another popular strategy for this number sentence is to adjust 268 to make it a compatible multiple of 5, then adjust the sum accordingly. In this way, 425 + 268 becomes 425 + 265 + 3.

3. Discuss the number sentences at the bottom of the OHT. For each example, call upon individuals to say the **partial sums** as they add. Examples 7, 8 and 9 may be more challenging, as these require the students to bridge over a hundred. Allow time for the students to complete Activity 1.2 in their Mastery Booklet.

Adding left to right

An ultralight plane flew at an altitude of 425 feet. After it climbed another 268 feet, what was its new altitude?

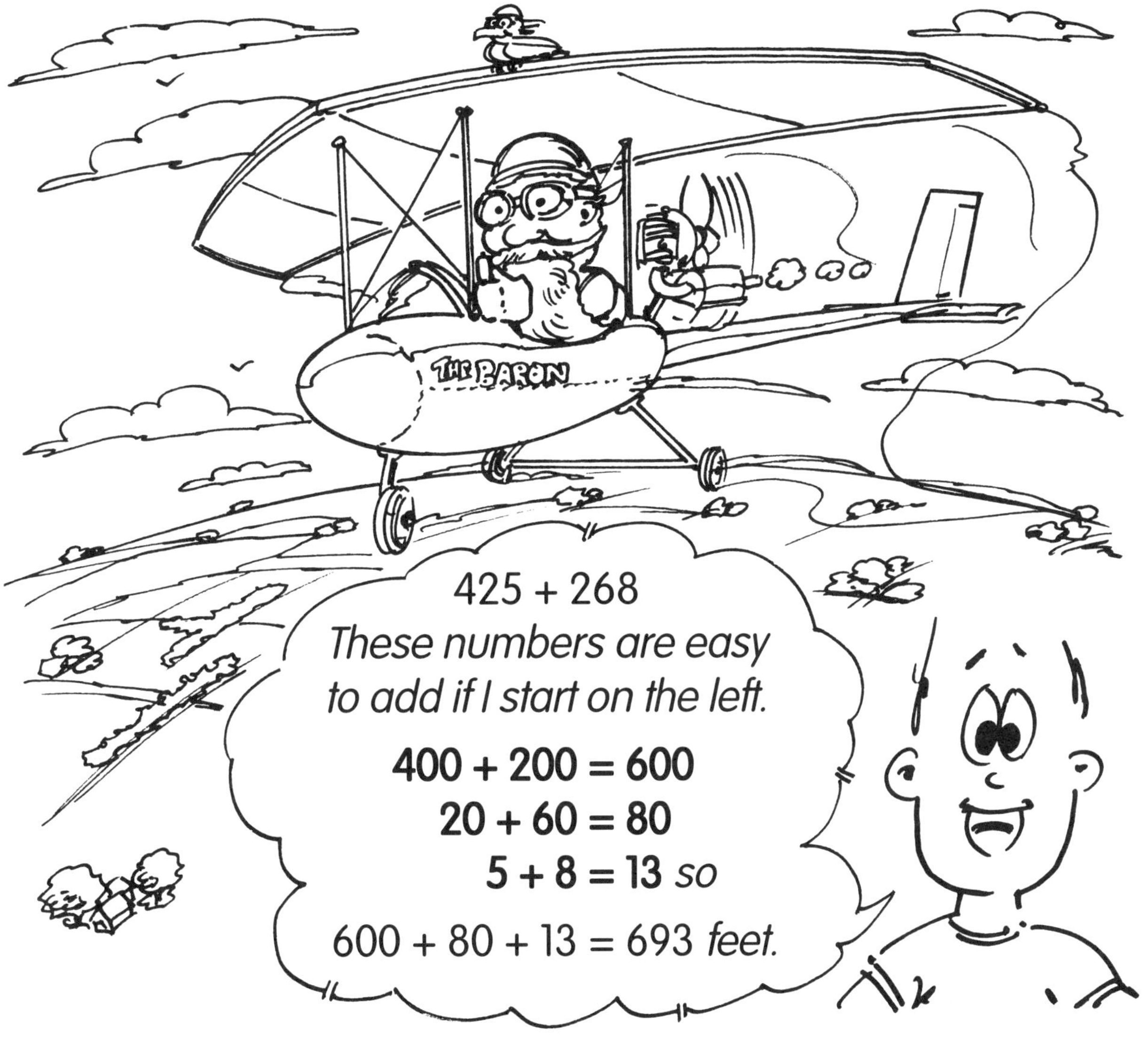

THE BARON

425 + 268

These numbers are easy to add if I start on the left.

400 + 200 = 600

20 + 60 = 80

5 + 8 = 13 *so*

600 + 80 + 13 = 693 *feet.*

Calculate the answers to these by adding the left first.

1. **245 + 137**
2. **426 + 258**
3. **317 + 246**
4. **327 + 245**
5. **638 + 134**
6. **513 + 428**
7. **345 + 273**
8. **284 + 345**
9. **168 + 451**

Adding decimals involving tenths

Introduction

In this activity, the students use a 'left-to-right' strategy to add decimal fractions that may or may not require bridging from tenths to ones. For example, 45.3 + 23.6 is calculated by adding (40 + 20) + (5 + 3) + (0.3 + 0.6).

Procedure

1. Read the problem at the top of OHT 1.3 together with the students. Challenge them to calculate the answer to the problem in their heads. Invite volunteers to share the strategies they used. For example, some students may prefer to ignore the decimal point when adding. In this way, 45.3 + 23.6 is the same as 453 tenths + 236 tenths or 68.9. Others may suggest a 'breaking-up' strategy, that is 45.3 + 20 + 3 + 0.6.

2. Read the strategy described in the speech balloon, then discuss the number sentences at the bottom of the OHT. For each example, call upon individuals to say the partial sums as they add. Alternatively, you may like to calculate the total of a few examples by ignoring the decimal point and treating the addends as tenths only. Examples 7, 8 and 9 may be more challenging, as these require the students to bridge from tenths to ones when using a 'left-to-right' strategy. You may want to suggest a different strategy for these number sentences to avoid the problems associated with bridging. For example, the addends could be adjusted so 34.6 + 24.7 becomes 35 + 24.3. Remember, there is no one correct strategy. Thinking such as this should be praised and encouraged.

3. Allow time for the students to complete Activity 1.3 in their Mastery Booklet.

Adding left to right

We had 45.3 mm of rain on Saturday and 23.6 mm on Sunday. How many millimeters of rain did we receive on the weekend?

68 ones

45.3 + 23.6

9 tenths

That's 68.9 mm *of rain.*

Calculate the answers to these by adding the left first.

1. **25.5 + 14.3**	4. **46.3 + 23.5**	7. **34.6 + 24.7**
2. **36.4 + 12.5**	5. **15.6 + 32.3**	8. **64.3 + 23.8**
3. **37.2 + 21.5**	6. **25.7 + 14.2**	9. **51.7 + 24.5**

Adding dollars and cents

Introduction

In this activity, the students use a 'left-to-right' strategy to add dollar amounts such as \$15.35 + \$4.24. Examples such as this do not require bridging. The students will also be encouraged to discuss other strategies for calculating number sentences involving dollars and cents.

Procedure

1. Show and read the problem at the top of OHT 1.4 to the students and challenge them to calculate the answer mentally.

2. Read the strategy discussed in the first speech balloon. Call upon an individual to verbalize the strategy in his/her own words. Briefly survey the class to determine who used this method, then read the second speech balloon to begin a discussion of other strategies the students may have used. For example, some students may break the second addend into its parts, starting with \$15.35 and adding \$4.00 + 20 cents + 4 cents. Others may prefer to follow a traditional paper-and-pencil algorithm, adding from the right first.

3. Discuss the number sentences at the bottom of the OHT. For each example, invite volunteers to explain the strategy they would use to calculate the sum. Examples 7, 8 and 9 may be more challenging, as these require bridging. You may want to suggest a different strategy for these number sentences to avoid the problems associated with bridging. For example, the addends could be adjusted so \$5.66 + \$13.28 becomes \$5.70 + \$13.24. Alternatively, the addends could be rounded and the sum adjusted accordingly. In this way, \$5.66 + \$13.28 becomes \$5.70 + \$13.30 – 6 cents. These strategies are investigated in Unit 4. Remember, there is no one correct strategy. Thinking such as this should be praised and encouraged.

4. Allow time for the students to complete Activity 1.4 in their Mastery Booklet.

Adding dollars and cents

Dad paid $15.35 for fuel and $4.24 for a can of oil.
How much did he spend altogether?

PLEASE PAY HERE

I'll add the dollars then the cents.

19 dollars

$15.35 + $4.24

59 cents

That's $19.59

What strategy did you use?

Calculate the answers to these in your head.

1. **$12.35 + $3.44**	4. **$15.52 + $3.34**	7. **$5.66 + $13.28**
2. **$2.12 + $5.47**	5. **$34.35 + $2.43**	8. **$6.47 + $12.34**
3. **$23.24 + $4.53**	6. **$22.61 + $4.35**	9. **$65.56 + $3.35**

Review

Procedure

1. Read the top of OHT 1.5 to the students. Ensure the remainder of the OHT is covered. Call upon a volunteer to mentally calculate the answer to Example A and then explain his/her thinking. Repeat this for the other three problems.

2. Show the students the second part of the OHT. Ask the students to calculate the answer to Example E in their heads. Reveal the first speech balloon to lead a discussion about the strategy they used. For example, some students may use a 'breaking-up' strategy, starting with $26.38 then adding $2.00 + 50 cents + 7 cents. Others may adjust the addends so that $26.38 + $2.57 became $26.40 + $2.55. Another popular strategy is to round and adjust the sum. In this way, $26.38 + $2.57 becomes $26.40 + $2.60 – 5 cents. Encourage the students to share why they adopted their strategy for this particular example. Reveal the second speech balloon to challenge the students to give other number sentences they could solve using the same strategy.

3. Show the students the remainder of the OHT. For each example, ask volunteers to explain how they would calculate the answer mentally. You may want to explain the mental strategy you would use.

Review

Explain how you calculate the answer to each of these in your head.

a) **426 + 38**

b) **254 + 435**

c) **24.3 + 35.5**

d) **32.6 + 16.8**

Calculate this number sentence in your head.

e) **$26.38 + $2.57**

Calculate the answer to each of these in your head, then explain the strategy you used.

1. **264 + 28**
2. **23.4 + 16.3**
3. **324 + 163**
4. **$12.62 + $3.25**
5. **42.5 + 23.7**
6. **$32.33 + $4.55**
7. **426 + 325**
8. **21.2 + 17.9**
9. **$6.28 + $3.37**
10. **235 + 46**

Breaking up three-digit addends

Introduction

One easy way to mentally calculate number sentences such as 125 + 143 is to break one **addend** into its parts and then add the parts separately. In this way, 125 + 143 becomes 125 + 100 + 40 + 3. The examples in this activity do not require bridging. This is an important prerequisite to Activity 2.2.

Procedure

1. Read the problem at the top of OHT 2.1 together with the students. Challenge them to calculate the **sum** in their heads. Ask those who were able to find the answer to share their strategies. Try to identify and discuss as many strategies as possible. The students need to see that there is often more than one way to add numbers mentally. For example, some students may use the 'left-to-right' strategy that was investigated in the previous unit. Others may follow a traditional written algorithm in their heads, adding the right first.

2. Say, *Sometimes it is easy to add in our heads by breaking a number into its parts.* Show and read the strategy described in the thought balloon. Call upon an individual to verbalize the strategy in his/her own words. For example, the child might say, *One hundred twenty-five, plus one hundred is two hundred twenty-five, plus forty is two hundred sixty-five, plus three is two hundred sixty-eight.* Then ask, *Could we break apart the other number instead?* (Yes). Invite a volunteer to verbalize how to calculate the sum by breaking the first addend into its parts.

3. Discuss the number sentences at the bottom of the OHT. For each example, invite a child verbalize how he/she would use the strategy to add the two numbers.

4. Give the students a copy of Blackline Master 2 (page 57). Show them how to fold the page in half and in half again to make a small booklet. Allow time for the students to finish the ten examples for Activity 2.1

Breaking up a number

Springfield School has 125 boys and 143 girls. How many students are there altogether?

I'll break 143 *into parts and add like this.*

125 + 100 + 40 + 3

What's the total?

Break a number apart to help you calculate the answers to these in your head.

1. **216 + 143**	4. **314 + 625**	7. **517 + 340**
2. **152 + 324**	5. **172 + 116**	8. **306 + 283**
3. **253 + 244**	6. **241 + 335**	9. **421 + 105**

Breaking up three-digit addends – with bridging

Introduction

In this activity, the students use a 'breaking-up' strategy to calculate number sentences such as 158 + 274. The examples in this activity require the students to bridge across a ten and/or hundred.

Procedure

1. Show the problem at the top of OHT 2.2 to the students. Ask them to calculate the answer in their heads. Invite those students who use a 'breaking-up' strategy to share their thinking. Then call upon the students who calculate the answer a different way to explain their strategy. For example, some students may still prefer to follow a paper-and-pencil algorithm in their heads. Others may use a 'left-to-right' strategy. For example, 158 + 274 is the same as (100 + 200) + (50 + 70) + (8 + 4). Alternatively, they may mentally add 15 tens + 27 tens + 12 ones.

2. Read the strategy shown in the speech balloon. Call upon an individual to verbalize the strategy in his/her own words. For example, the child might say, *One hundred fifty-eight, plus two hundred is three hundred fifty-eight, plus seventy is four hundred twenty-eight, plus four is four hundred thirty-two.* Show and discuss the number sentences at the bottom of the OHT. Encourage the students to say the parts of the second addend as they add. Examples 7, 8 and 9 may be more challenging, as these require bridging across a ten and hundred.

3. Allow time for the students to complete Activity 2.2 in their Mastery Booklet.

Breaking up a number

At the local show, 158 hotdogs were sold in the morning and 274 were sold in the afternoon. How many hotdogs were sold altogether?

Mmm … 158 + 274
is the same as
158 + 200 + 70 + 4
That's 432

Use the same thinking to calculate the answers to these in your head.

1. **468 + 324**	4. **347 + 426**	7. **356 + 275**
2. **253 + 138**	5. **284 + 342**	8. **467 + 354**
3. **325 + 246**	6. **462 + 485**	9. **187 + 444**

Breaking up two addends – with bridging

Introduction

A 'breaking-up' strategy is a natural way of mentally adding number sentences such as 126 + 33 + 14. Children, who prefer to follow a paper-and-pencil strategy in their heads, may find that it is an inefficient method for adding more than two numbers — particularly when bridging is involved. In this activity, the students break two addends into parts to help them find the sum of three numbers.

Procedure

1. Show the top of OHT 2.3 and invite a volunteer to read the problem to the class. Ask the students to share how they would calculate the answer in their heads. At this stage, the students should know they could use a 'breaking-up' strategy to calculate the answer. However, some students may prefer a different method. For example, they may add left to right — 12 tens + 3 tens + 1 ten = 16 tens, and 6 + 3 + 4 = 13 ones. 160 + 13 = 173. The students should be encouraged to use a strategy that best suits their way of thinking.

2. Read the strategy described in the speech balloon. Call upon a volunteer to explain the strategy in his/her own words. The child should say that this method requires them to start with 126 and add each part of second addend, then each part of the last addend. Ask the students if they could use this strategy another way. Allow time for them to share their thinking. Encourage them to see that any two numbers could be broken into parts, however it is easier to begin with the greatest number and add the parts of the other two numbers.

3. Show and discuss the examples at the bottom of the OHT. Encourage the students to say the parts of the addends as they add. For Examples 4 to 9, the students should start with the greatest number then add the parts of the other two numbers. Examples 7, 8 and 9 may be more challenging, as these require the students to bridge over a ten and hundred.

4. Allow time for the students to complete Activity 2.3 in their Mastery Booklet.

Breaking up two numbers

A plane has 126 seats in economy class, 33 in business class and 14 in first class. How many seats are there in all?

126 + 33 + 14

Try breaking up two numbers.

126 + 30 + 3 + 10 + 4 = 173

That's 173 *seats!*

Break up two numbers in each of these examples to help you calculate the sum in your head.

1. **216 + 33 + 46**
2. **134 + 27 + 24**
3. **348 + 35 + 16**
4. **26 + 135 + 37**
5. **28 + 24 + 247**
6. **31 + 224 + 28**
7. **446 + 35 + 42**
8. **38 + 255 + 43**
9. **54 + 38 + 343**

Breaking up two decimals involving tenths – with bridging

Introduction

In this activity, the students use a 'breaking-up' strategy to calculate number sentences such as 21.4 + 2.7 + 4.6. The examples in this activity require the students to bridge from tenths to ones and/or across a ten.

Procedure

1. Show OHT 2.4 and invite a volunteer to read aloud the problem at the top. Some students may be able to mentally compute the answer using a method other than the one that is discussed on the OHT. Ask the students to calculate the answer in their heads, then encourage them to share their methods. Some students may suggest using a 'left-to-right' strategy to obtain the sum. Using this method the students would first add 21 + 2 + 4 then 0.4 + 0.7 + 0.6 before combining the two partial sums. Another popular strategy is to use a compatible pair. For example, 21.4 and 4.6 add to make a whole number. In this case, the students think, *Twenty-six plus two and seven tenths is twenty-eight and seven tenths.*

2. Discuss the summary of the strategy shown in the speech balloon. Call upon an individual to verbalize the strategy in his/her own words. In this case, the students should say,

   Twenty-one and four tenths,
   plus two is twenty-three and four tenths,
   plus seven tenths is twenty-four and one tenth,
   plus four is twenty-eight and one tenth,
   plus six tenths is twenty-eight and seven tenths.

   Ask the students how this method compares to the methods shared in the discussion above. *Is there an easier way to add these three numbers?* Invite the students to share their thinking.

3. Discuss the number sentences at the bottom of the OHT. Encourage the students to say the parts of the addends as they add. Examples 4, 5 and 6 may be more challenging than the first three examples, as these require the students to bridge from tenths to ones and across a ten.

4. Allow time for the students to complete Activity 2.4 in their Mastery Booklet.

Breaking up two numbers

We boarded the train and got off three stations later. The stations were 21.4 km, 2.7 km and 4.6 km apart. How far did we travel altogether?

I'll add the parts like this.

21.4 + 2.7 + 4.6

What's the total?

Use the same thinking to calculate the answers to these in your head.

1. **23.2 + 3.4 + 2.5**
2. **4.2 + 13.7 + 1.8**
3. **3.1 + 2.5 + 22.7**
4. **33.4 + 5.5 + 6.2**
5. **4.8 + 28.4 + 3.1**
6. **5.2 + 17.3 + 7.5**

Review

Procedure

1. Read the sentence at the top of OHT 2.5 to the students. Ensure the remainder of the OHT is covered. Allow time for the students to calculate the answers to Examples A, B, C and D. For each example, encourage them to share their thinking, then ask them to suggest other number sentences they could solve using the same strategy.

2. Show the students the second part of the OHT. Ask them to calculate the answer to Example E in their heads. Read the speech balloons to begin a discussion about their strategies. Some students may use a 'breaking-up' or a 'left-to-right' strategy. Others may choose to add a compatible pair first. For example, 23.7 and 6.3 add to make a whole number. In this case the students think, *Thirty plus two and five tenths is thirty-two and five tenths.* Afterwards, ask the students to share why they adopted their strategy for this particular example.

3. Discuss the number sentences at the bottom of the OHT. For each example, ask volunteers to explain how they would calculate the answer mentally. Remember, no one strategy is the correct strategy. Allow the students to use the strategy that best suits their way of thinking. Flexible thinking should be praised and encouraged.

Review

Explain how you calculate the answer to each of these in your head.

a) **234 + 145**

b) **356 + 275**

c) **324 + 18 + 35**

d) **48 + 164 + 23**

Calculate this number sentence in your head.

e) **23.7 + 2.5 + 6.3**

Calculate the answer to each of these in your head.

1. **346 + 223**
2. **235 + 23 + 28**
3. **257 + 138**
4. **31.5 + 3.7 + 2.6**
5. **42 + 119 + 27**
6. **362 + 259**
7. **4.5 + 2.7 + 12.4**
8. **35 + 46 + 271**
9. **6.3 + 4.8 + 32.7**
10. **624 + 155**

Breaking up dollars and cents

Introduction

In this activity, the students use a 'breaking-up' strategy to calculate number sentences involving tenths and hundredths. For example, \$4.35 + \$2.50 is the same as \$4.35 + \$2.00 + 50 cents. The number sentences in this activity do not require bridging and include one addend that is a multiple of 10 cents.

Procedure

1. Show and read the problem at the top of OHT 3.1 to the students. Make sure the remainder of the OHT is covered. Ask them to calculate the answer in their heads. Invite two or three individuals to share and explain their strategies. For example, some students may use a 'left-to-right' strategy, adding \$4.00 + \$2.00 then 35 cents + 50 cents. Others may use the 'breaking-up' strategy described in the speech balloon.

2. Call upon a volunteer to read the strategy in the speech balloon. You may want to model the strategy using play money. Count out the first amount (\$4.35) and place it in the hand of a volunteer. Then give the second amount (a \$2.00 coin and a 50 cent coin) to the child adding the parts aloud. For example, you should say, *Four dollars thirty-five, plus two dollars is six dollars thirty-five, plus fifty cents is six dollars eighty-five.*

3. Discuss the number sentences at the bottom of the OHT. For the first few examples, call upon individuals to add the parts aloud. Allow the students to use the play money if they wish.

4. Give the students a copy of Blackline Master 3 (page 58). Show them how to fold the page in half and in half again to make a small booklet. Allow time for the students to finish the ten examples for Activity 3.1.

Breaking up dollars and cents

Burgers are $4.35 and drinks are $2.50. How much will it cost to buy one of each?

$4.35 + $2.50
is the same as
$4.35 + $2.00 + 50 cents
What's the total?

Use the same strategy to add these in your head.

1. **$3.45 + $1.40**	4. **$6.37 + $1.40**	7. **$5.40 + $3.47**
2. **$6.25 + $2.70**	5. **$5.18 + $3.50**	8. **$2.30 + $5.38**
3. **$2.15 + $4.60**	6. **$4.46 + $2.30**	9. **$1.10 + $3.76**

Breaking up dollars and cents

Introduction

In this activity, the students use a 'breaking-up' strategy to calculate amounts such as $13.45 + $5.34. The number sentences in this activity do not require bridging.

Procedure

1. Show and read the problem at the top of OHT 3.2 to the students. Ask the students how these prices compare to compact discs or cassettes they may have bought. Encourage them to suggest reasons for any price differences. For example, they might say these items could be old stock, second hand or artist singles.

2. Challenge the students to calculate the sum in their heads. Call upon those who obtain the correct answer to share their strategies. Some students may use a 'breaking-up' or 'left-to-right' strategy. Another popular method is to round and adjust the answer accordingly. For example, $13.45 + $5.34 is the same as $13.50 + $5.30 – 1 cent. Thinking such as this should be praised and encouraged.

3. Reveal and read the thought balloon to the class, then discuss the number sentences at the bottom of the OHT. Encourage the students to use the strategy that best suits their way of thinking. For Examples 4, 5, 6 and 8, ask the students which amount they would break up and why. The students should say they would start with the greater amount and break up the smaller amount.

4. Allow time for the students to complete Activity 3.2 in their Mastery Booklet.

Breaking up dollars and cents

Karl bought a CD for $13.45 and a cassette for $5.34. What was the total cost?

Mmm ...

$13.45 + $5.00 + 34 cents

That's $18.79

Calculate the answer to each of these.

1. **$24.15 + $4.53**	4. **$5.45 + $13.33**	7. **$42.21 + $3.54**
2. **$33.25 + $5.34**	5. **$6.41 + $22.25**	8. **$4.63 + $31.24**
3. **$21.75 + $3.22**	6. **$4.35 + $34.13**	9. **$32.16 + $3.62**

Breaking up dollars and cents – with bridging

Introduction

In this activity, the students use a 'breaking-up' strategy to calculate amounts such as \$22.35 + \$4.48. The number sentences in this activity require the students to bridge from hundredths to tenths of a dollar and/or across a dollar.

Procedure

1. Show the problem at the top of OHT 3.3 to the students. Make sure the remainder of the OHT is covered. Invite a volunteer to read the problem, calculate the sum mentally and share his/her answer with the class. If the volunteer can tell you the correct answer, call upon several students to guess the strategy that was used. Encourage the class to suggest as many different strategies as possible before asking the volunteer to explain his/her strategy. The aim of this discussion is to make the students aware of the many strategies they could use to solve this particular problem. For example, in addition to a 'breaking-up' and 'left-to-right' strategy, the students could round and adjust the answer accordingly. That is, \$22.35 + \$4.48 is the same as \$22.40 + \$4.50 – 7 cents. Another popular strategy is to adjust the addends to make one a multiple of 10 cents. In this way, \$22.35 + \$4.48 becomes \$22.40 + \$4.43. These two methods were explored in Activity 3.3 and 3.4 of the addition book, *Thinking MatheMentally: Whole Numbers and Decimals* and are further investigated in Activity 4.2 and 4.3 of this book.

2. Reveal and read the strategy described in the speech balloon. Invite an individual to verbalize the strategy in his/her own words.

3. Discuss the number sentences at the bottom of the OHT. Encourage the students to use any one of the strategies that were previously mentioned. For Examples 4, 5, 6 and 9, ask the students which amount they would break up and why. The students should say they would start with the greater amount and break up the smaller amount. Examples 7, 8 and 9 may be more challenging, as these require bridging from hundredths to tenths of a dollar and from tenths to a whole dollar.

4. Allow time for the students to complete Activity 3.3 in their Mastery Booklet.

Breaking up dollars and cents

Nicole's savings earned $22.35 interest in the first half of the year and $4.48 in the second half. How much interest was earned altogether?

BANK PASS BOOK

Mmm ... $22.35 + $4.48

$22.35 + **$4.00** = $26.35

$26.35 + **40 cents** = $26.75

$26.75 + **8 cents** = $26.83

Calculate the sum of each of these in your head.

1. **$31.45 + $2.37**
2. **$22.65 + $3.29**
3. **$15.25 + $1.68**
4. **$4.37 + $22.54**
5. **$5.62 + $31.19**
6. **$3.24 + $12.47**
7. **$45.36 + $2.75**
8. **$23.48 + $4.85**
9. **$5.74 + $21.88**

Breaking up dollars and cents – with bridging

Introduction

In this activity, the students use a 'breaking-up' strategy to calculate amounts such as $8.50 + 85 cents. The number sentences in this activity have one addend less than $1.00. These also require the students to bridge from hundredths to tenths of a dollar and/or across a dollar.

Procedure

1. Show the illustration and read the problem at the top of OHT 3.4 to the students. Ensure the thought balloon remains covered. Challenge the students to calculate the sum in their heads. Call upon individuals to share and explain their strategies. Some students may suggest a 'breaking-up' strategy, but there are at least two other popular methods that could be applied to this particular number sentence. For example, the amounts could be adjusted to make one a whole dollar. In this way, $8.50 + 85 cents becomes $9.00 + 35 cents, or $8.35 + $1.00. Another method is to round the smaller amount and adjust the answer accordingly, so $8.50 + 85 cents becomes $8.50 + $1.00 – 15 cents.

2. Reveal and read the strategy described in the thought balloon. Find out the number of students who used this strategy. Call upon one or two of these students to explain why they chose this particular strategy for the number sentence. Remember there is no one correct strategy to use. The aim is to help the students understand that they need to think about the full range of strategies before choosing an appropriate method to use.

3. Show and discuss the number sentences at the bottom of the OHT. For each example, invite volunteers to identify and explain a mental strategy that they feel is appropriate. For Examples 4, 5, 6 and 8, ask the students which amount they would break up and why. The students should say they would start with the greater amount and break up the smaller amount. Examples 7, 8 and 9 may be more challenging, as these require bridging from hundredths to tenths of a dollar and across a dollar.

4. Allow time for the students to complete Activity 3.4 in their Mastery Booklet.

Breaking up dollars and cents

Jesse needs new batteries for her radio. A package of batteries costs \$8.50 plus 85 cents tax. How much will she spend altogether?

BATTERIES

I'll add the parts of the smaller amount.

\$8.50 + **80 cents** = \$9.30

\$9.30 + **5 cents** = \$9.35

Calculate the sum of each of these in your head.

1. **\$8.90 + 89¢**
2. **\$7.80 + 78¢**
3. **\$8.60 + 86¢**
4. **85¢ + \$12.60**
5. **45¢ + \$45.70**
6. **49¢ + \$33.90**
7. **\$16.35 + 89¢**
8. **48¢ + \$26.87**
9. **\$36.24 + 87¢**

Review

Procedure

1. Read the sentence at the top of OHT 3.5 to the students. Ensure the remainder of the OHT is covered. Allow time for the students to calculate the answers. For each example, encourage them to share their thinking, then ask them to suggest other number sentences they could solve using the same strategy. Point to Example D. Call upon those students who used a 'breaking-up' strategy to explain which addend they chose to break into parts and why. The discussion will vary, but they should say they start with the greater amount ($13.55) and break up the smaller amount ($3.68).

2. Show the students the second part of the OHT. Ask them to calculate the answer to Example E in their heads. Read the speech balloons to begin a discussion about the strategy they used. Some students may use a 'breaking-up' strategy. Others may decide to adjust the amounts to make one a whole dollar, for example, $23.90 + 48 cents is the same as $24.00 + 38 cents. Another popular strategy to round one or both amounts and adjust the sum accordingly. In this way, $23.90 + 48 cents becomes $23.90 + 50 cents – 2 cents, or $24.00 + 50 cents – 12 cents. Afterwards, encourage the students to share why they adopted their strategy for this particular example.

3. Discuss the number sentences at the bottom of the OHT. For each example, ask volunteers to explain how they would calculate the answer mentally. At this stage you may want to share the strategy you would use.

Review

Explain how you calculate the sum of each of these in your head.

a) **\$5.60 + \$2.27**

b) **\$22.35 + \$6.42**

c) **\$14.25 + \$4.37**

d) **\$3.68 + \$13.55**

Calculate this number sentence in your head.

e) **\$23.90 + 48¢**

Calculate the answer to each of these in your head, then explain the strategy you used.

1. **\$3.40 + \$2.55**
2. **\$4.42 + \$5.59**
3. **\$2.42 + \$5.37**
4. **\$7.40 + 85¢**
5. **\$23.68 + \$4.17**
6. **\$6.23 + \$1.54**
7. **48¢ + \$8.78**
8. **\$4.74 + \$13.88**
9. **\$6.25 + \$3.60**
10. **\$16.45 + 79¢**

Rounding and adjusting three-digit numbers

Introduction

In this activity, the students use rounding and adjusting to calculate number sentences such as 296 + 148. One way to add numbers such as these is to adjust the addends to make one a multiple of 10, so 296 + 148 becomes 300 + 144, or 294 + 150. Another popular method is to round one or both addends to a multiple of 10 then adjust the answer, for example 296 + 148 is the same as 300 + 148 – 4, 300 + 150 – 6. Unit 3 of the addition book, *Thinking MatheMentally: Whole Numbers and Decimals* is the prerequisite for this unit.

Procedure

1. Show and read the top of OHT 4.1. Represent 294 and 148 using base-ten blocks, then ask the students how they would add the two numbers in their heads. Elicit several different responses. For example, some students may suggest using a 'left-to-right' strategy, adding the place values of each number separately. In this way, 296 + 148 is calculated as (200 + 100) + (90 + 40) + (6 + 8). Others may suggest using a 'breaking-up' strategy starting with 296 then adding 100, 40 then 8. Without discouraging the use of these strategies, you may want to point out the number of steps that are involved and the fact that they both require a great deal of bridging. If a 'rounding-and-adjusting' strategy is not suggested ask, *How can we change these two numbers so they are easier to add in our heads?* Invite volunteers to share their thinking. You may need to model the idea of taking four 'ones' blocks from 148 and adding it to 296. If so ask, *Has the total number of blocks changed?* (No). *What are the new numbers?* (300 and 144). *Are these numbers easier to add in our heads? Could we change the numbers another way to make the numbers even easier to add?* Allow the students to share their opinions.

2. Show and read the strategies described in the speech balloons. Check that the students understand the second strategy and the idea of rounding the numbers. Discuss the number sentences at the bottom of the OHT. For each example, call upon a volunteer to verbalize the strategy they would use. Remember there is no one correct strategy. The students should be encouraged to use the strategy that best suits their way of thinking.

3. Give the students a copy of Blackline Master 4 (page 59). Show them how to fold the page in half and in half again to make a small booklet. Allow time for the students to finish the ten examples for Activity 4.1.

Rounding and adjusting numbers

The Johnsons drove 296 miles on Saturday and another 148 miles on Sunday to reach their destination. How far did they travel?

THE SEA 148

The answer is a little more than 440
296 + 148
is the same as
300 + 144

296 +148
is also the same as
300 + 150 – 6
That's 444 *miles!*

Try using one of the methods above to calculate the answer to each of these.

1. **398 + 296**
2. **249 + 397**
3. **499 + 148**
4. **154 + 196**
5. **347 + 254**
6. **153 + 298**
7. **548 + 354**
8. **695 + 246**
9. **357 + 245**

Rounding and adjusting decimals involving tenths

Introduction

In this activity, the students use rounding and adjusting to calculate number sentences such as 41.8 + 25.7. The first strategy involves adjusting the addends, so 41.8 + 25.7 becomes 42 + 25.5. Another popular method is to round one or both addends to a whole number then adjust the answer, for example 41.8 + 25.7 is the same as 42 + 26 – 0.5. These strategies eliminate the need to bridge.

Procedure

1. Show the illustration and read the problem at the top of OHT 4.2. Ensure the remainder of the OHT is covered. Ask the students how they would calculate 41.8 + 25.7 in their heads. Call upon several students to share their thinking. Some students may suggest the 'left-to-right' strategy investigated in Unit 1. Others may want to break up the second addend, starting with 41.8 and adding 25 then 0.7. If an 'adjusting' strategy is not suggested ask, *How can we change these two numbers so they are easy to add in our heads?* Elicit several responses from the children, then show and read the strategy described in the first speech balloon. Check they understand that 2 tenths were taken from the second addend and added to the first.

2. Reveal and read the second speech balloon to the students. Ask, *Could we use rounding another way?* The students should see that it is not necessary to round both addends.

3. Discuss the number sentences at the bottom of the OHT. For each example, call upon a volunteer to verbalize the strategy they would use. Remember that no one strategy is the correct one. The students should be encouraged to use the strategy that best suits their way of thinking.

4. Allow time for the students to complete Activity 4.2 in their Mastery Booklet.

Rounding and adjusting decimals

Capetown is 25.7 miles north west of Middletown and Middletown is 41.8 miles north west of Forest Ridge. How far is Capetown from Forest Ridge?

CAPETOWN
25.7
MIDDLETOWN
41.8
FOREST RIDGE
N
W
E
S

The answer is about 68.
41.8 + 25.7
is the same as
42 + 25.5

41.8 + 25.7
is also the same as
42 + 26 – 0.5
That's 67.5 *miles!*

Use one of the methods above to calculate the answer to each of these.

1. **21.8 + 32.7**
2. **24.9 + 12.8**
3. **35.7 + 13.6**
4. **14.7 + 24.8**
5. **25.6 + 12.9**
6. **36.8 + 11.8**
7. **41.7 + 24.7**
8. **35.9 + 23.9**
9. **17.8 + 21.9**

Rounding and adjusting dollars and cents

Introduction

In this activity, the students investigate two strategies for adding dollar amounts such as $10.95 + $14.95. The first strategy involves adjusting the addends so $10.95 + $14.95 becomes $11.00 + $14.90. The second method involves rounding and adjusting the answer, for example $10.95 + $14.95 is the same as $11.00 + $15.00 – 10 cents.

Procedure

1. Read the problem at the top of OHT 4.3 to the students. Ask them how they would calculate the sum in their heads. Encourage the students to share their thinking.

2. Read the strategies shown in the speech balloons. Check that the students understand the first strategy. Ask them how $10.95 and $14.95 were changed to $11.00 and $14.90 (five cents was taken from the second amount and added to the first). Briefly survey the class to identify the students who used one of the strategies described on the OHT. The flexible thinking required for strategies such as these should be praised.

3. Discuss the number sentences on the bottom of the OHT. Encourage the students to use one of the described strategies to help them calculate the answers.

4. Allow time for the students to complete Activity 4.3 in their Mastery Booklet.

Rounding and adjusting dollars and cents

Michelle bought a CD cleaner for $10.95 and a CD storage rack for $14.95. How much did she spend?

I can do this two ways!
$10.95 + $14.95
is the same as
$11.00 + $14.90

$10.95 + $14.95
is also the same as
$11 + $15 – 10 cents
That's $25.90

Use one of the methods above to calculate the answer to each of these.

1. **$13.95 + $25.95**
2. **$24.97 + $14.98**
3. **$32.97 + $24.97**
4. **$18.98 + $17.98**
5. **$26.99 + $17.95**
6. **$42.98 + $17.99**

Rounding and adjusting dollars and cents

Introduction

In this activity, the students extend and apply the strategies that were investigated in Activity 4.2 to dollar amounts such as $23.48 + $16.47. The first strategy involves adjusting the addends so $23.48 + $16.47 becomes $23.50 + $16.45. The second method involves rounding and adjusting the answer, for example $23.48 + $16.47 is the same as $23.50 + $16.50 – 5 cents.

Procedure

1. Show and read the problem at the top of OHT 4.4 to the students. Challenge them to calculate the answer in their heads. Discuss the strategy shown in the first speech balloon. Survey the class to determine how many students used the same strategy. Then ask, *How was the new number sentence* ($23.50 + $16.45) *made?* Make sure the students understand that 2 cents was taken from $16.47 and added to $23.48 to make $23.50. *Who used the same strategy a different way?* Some students may choose to take 3 cents from $23.48 and add it to $16.47 to make $16.50. Encourage these students to share and explain their strategy.

2. Read aloud the second speech balloon to begin a discussion about other strategies that could be used to calculate the total. Adding 'left-to-right' is a popular strategy for number sentences such as this. This method requires the students to think ($23 + $16) + (48 cents + 47 cents). Make sure a 'rounding-and-adjusting' strategy is mentioned. This method requires the students to round one or both amounts to the nearest ten cents before adding them and adjusting the answer. In this way $23.48 + $16.47 becomes $23.50 + $16.50 – 5 cents.

3. Discuss the number sentences at the bottom of the OHT. For each example, have the students tell you an approximate answer before calculating the exact sum, then encourage volunteers to share their methods. The students could use one of several strategies for Example 6. Many students will automatically know double 45 cents is 90 cents, so they may use a 'left-to-right' strategy to calculate the answer.

4. Allow time for the students to complete Activity 4.4 in their Mastery Booklet.

Rounding and adjusting dollars and cents

Last month, Janelle spent $23.48 on local calls and $16.47 on long distance calls. What was the total cost of her phone calls?

PHONE BILL

The total is about $40.
$23.48 + $16.47
is the same as
$23.50 + $16.45
That's $39.95

Describe another strategy you could use.

Calculate the sum of each of these in your head.

1. **$23.48 + $15.47**
2. **$34.45 + $21.49**
3. **$36.49 + $12.48**
4. **$16.47 + $31.45**
5. **$20.48 + $13.48**
6. **$14.45 + $24.45**

Review

Procedure

1. Briefly discuss the strategies that were investigated in the previous activities. Read the top of OHT 4.5 and show Examples A, B, C and D. Ensure the remainder of the OHT is covered. Challenge the students to calculate the answers, then call upon individuals to share and explain their strategies.

2. Show the students the second part of the OHT. Ask the students to calculate the answer to Example E in their heads. Lead a discussion about the thinking they used. Make a list of the strategies on the board. At this stage, the students should be aware of several possibilities. For example, $24.45 + $23.45 could be added in each of the following ways:

   $24.40 + $23.50
   $24.50 + $23.40
   $24.50 + $23.50 – 10 cents
   $47.00 + 90 cents
   $24.45 + $23.00 + 45 cents

   Invite students to justify the use of their strategy over the others in the list.

3. Show the students the remainder of the OHT. For each example, ask the students to calculate the answer and share the strategy they used. Remember that no one strategy is correct. Encourage the students to use the strategy that best suits their way of thinking. You may want to share the strategy you would use for each example.

Review

Tell how you calculate each of these number sentences in your head.

a) **297 + 348**

b) **23.7 + 12.8**

c) **$23.95 + $14.98**

d) **$12.48 + $24.49**

Calculate this number sentence in your head.

e) **$24.45 + $23.45**

Calculate the answer to each of these in your head, then explain the strategy you used.

1. **23.7 + 14.8**
2. **$23.48 + $14.48**
3. **$15.95 + $13.98**
4. **348 + 298**
5. **$13.45 + $12.49**
6. **197 + 248**
7. **$24.97 + $31.95**
8. **$21.48 + $22.47**
9. **149 + 354**
10. **15.9 + 33.8**

Adding fractions with like denominators

Introduction

In this activity, the students calculate number sentences involving fractions that have like **denominators**, such as $^3/_8 + ^4/_8$. This example does not involve bridging, however Examples 4 to 9 on OHT 5.1 require the students to bridge over one whole.

Procedure

1. Show the illustration and read the problem at the top of OHT 5.1 to the students. Make sure the remainder of the OHT is covered. Invite an individual to tell you the amount of pizza that is left over.

2. Reveal and read the strategy described in the thought balloon, then discuss Examples 1, 2 and 3 at the bottom of the OHT. These examples do not require bridging and should be easy to calculate quickly. Point to Example 4 and ask the students to calculate the sum in their heads. Call upon two or three students who obtain the correct answer to share their strategies. Some students may add $^7/_8 + ^6/_8$ and give the sum as an **improper fraction** ($^{13}/_8$) or convert this to a **mixed number** ($1^5/_8$). Others may apply one of the 'rounding-and-adjusting' strategies that were investigated in Unit 4. For example, $^7/_8 + ^6/_8$ is the same as $1 + ^5/_8$; or $1 + 1 - ^3/_8$. Discuss Examples 5 to 9. Encourage the students to use the strategy that best suits their way of thinking.

3. Give the students a copy of Blackline Master 5 (page 60). Show them how to fold the page in half and in half again to make a small booklet. Allow time for the students to complete the examples for Activity 5.1.

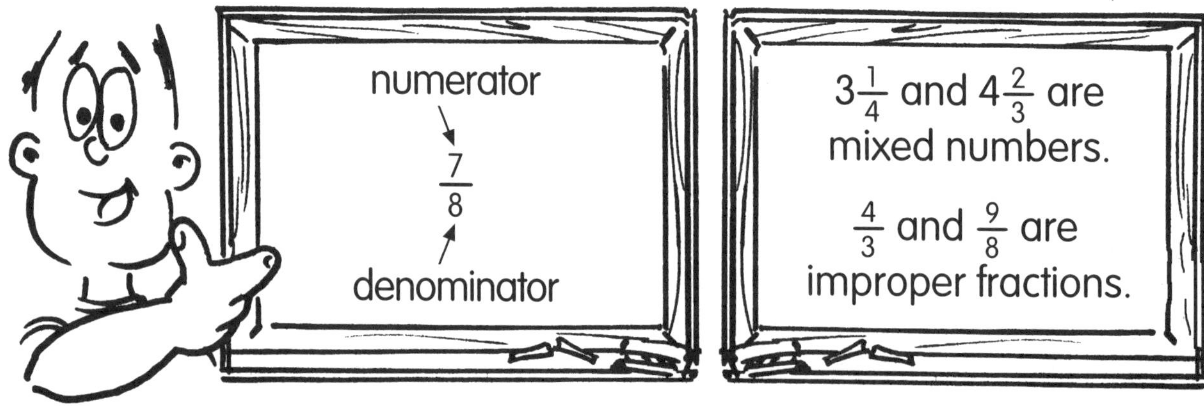

Adding fractions

We ordered several pizzas for our party. Afterwards we had $\frac{3}{8}$ of one pizza left over and $\frac{4}{8}$ of another left over. How much pizza was not eaten?

It's easy to add fractions when the denominators are the same.

3 eighths + 4 eighths = 7 eighths

$\frac{7}{8}$ *left over!*

Calculate the sum of each of these in your head.

1. $\frac{3}{5} + \frac{1}{5}$
2. $\frac{5}{8} + \frac{2}{8}$
3. $\frac{11}{16} + \frac{3}{16}$
4. $\frac{7}{8} + \frac{6}{8}$
5. $\frac{3}{4} + \frac{2}{4}$
6. $\frac{4}{5} + \frac{3}{5}$
7. $\frac{10}{16} + \frac{8}{16}$
8. $\frac{3}{8} + \frac{7}{8}$
9. $\frac{2}{5} + \frac{4}{5}$

Adding fractions with related denominators

Introduction

In this activity, the students calculate number sentences involving fractions that have related denominators, such as $^1/_2 + ^1/_4$. This problem does not involve bridging, however Examples 4 to 9 on OHT 5.2 require the students to bridge over one whole.

Procedure

1. In order to add fractions mentally, students need to be competent at identifying equivalent fractions. Write $^2/_8$ on the board. Ask the students to tell you two or three equivalent fractions. Call upon individuals to tell how they worked them out. Repeat this discussion with other fractions involving halves, thirds, quarters, eighths and sixteenths.

   Blackline Master 6 (page 61) provides further practice generating equivalent fractions. Give the students a copy of the sheet and allow time for them to complete the activity.

2. Show and read the problem at the top of OHT 5.2 to the students. Invite a volunteer to calculate the sum and explain the thinking that he/she used. The discussion will vary, but the child should say, *I change one half to two quarters then add one quarter.*

3. Reveal and read the strategy in the speech balloons, then discuss Examples 1, 2 and 3 at the bottom of the OHT. These examples do not require bridging and should be easy to calculate quickly. Point to Example 4 and ask the students to calculate the sum in their heads. Call upon two or three students who obtain the correct answer to share and explain their strategies. After finding the equivalent fraction, some students may add $^3/_4 + ^2/_4$ and give the sum as an improper fraction ($^5/_4$) or convert this to a mixed number ($1^1/_4$). Others may use a 'rounding-and-adjusting' strategy. For example, $^3/_4 + ^2/_4$ is the same as $1 + ^1/_4$. Discuss Examples 5 to 9. Encourage the students to use the strategy that best suits their way of thinking.

4. Allow time for the students to complete Activity 5.2 in their Mastery Booklet.

Adding fractions

Andrew mowed $\frac{1}{2}$ acre of grass in the morning and $\frac{1}{4}$ acre of grass in the afternoon. How much grass did he mow?

$\frac{1}{2} + \frac{1}{4}$

It's easy when one denominator is a multiple of the other.

$\frac{1}{2}$ *is the same as* $\frac{2}{4}$ *so*

$\frac{2}{4} + \frac{1}{4} = \frac{3}{4}$

Calculate the sum of each of these in our head.

1. $\frac{1}{2} + \frac{3}{8}$
2. $\frac{1}{4} + \frac{5}{8}$
3. $\frac{3}{8} + \frac{4}{16}$
4. $\frac{3}{4} + \frac{1}{2}$
5. $\frac{2}{4} + \frac{7}{8}$
6. $\frac{3}{8} + \frac{12}{16}$
7. $\frac{6}{8} + \frac{3}{4}$
8. $\frac{1}{2} + \frac{11}{16}$
9. $\frac{10}{16} + \frac{3}{4}$

Adding mixed numbers with like denominators

Introduction

In this activity, the students investigate strategies for solving number sentences involving mixed numbers with like denominators, such as $4^7/_8 + 3^6/_8$. The first strategy involves adjusting the fractions to make one a whole number, so $4^7/_8 + 3^6/_8$ becomes $5 + 3^5/_8$. The second method involves rounding one or both fractions then adjusting the sum accordingly, for example $4^7/_8 + 3^6/_8$ is the same as $5 + 4 - {}^3/_8$. These strategies eliminate the need to bridge.

Procedure

1. Show and read the problem at the top of OHT 5.3 to the students. Challenge them to calculate the answer in their heads. Discuss the strategy shown in the first speech balloon. Survey the class to determine how many students used this method. Then ask, *How was the new number sentence* ($5 + 3^5/_8$) *made?* Make sure the students understand that ${}^1/_8$ was taken from $3^6/_8$ and added to $4^7/_8$ to make 5. *Who used the same strategy a different way?* Some students may choose to take ${}^2/_8$ from $4^7/_8$ and add it to $3^6/_8$ to make 4. Encourage these students to share and explain their strategy.

2. Read aloud the second speech balloon to begin a discussion about other strategies that could be used to calculate the total. Some students may suggest using a 'left-to-right' strategy. For example, $4^7/_8 + 3^6/_8$ can be calculated by adding $(4 + 3) + ({}^7/_8 + {}^6/_8)$ which is the same as $7 + 1^5/_8$. Without discouraging the use of this strategy, you may want to point out the number of steps that are involved and the fact that it requires bridging. A popular strategy that does not requiring bridging, is rounding the addends and adjusting the answer. This method involves rounding one or both fractions to the nearest whole number before adding, then adjusting the answer. In this way, $4^7/_8 + 3^6/_8$ becomes $5 + 4 - {}^3/_8$.

3. Discuss the number sentences at the bottom of the OHT. For each example, invite volunteers to explain the methods they would use.

4. Allow time for the students to complete Activity 5.3 in their Mastery Booklet.

Adding fractions

A baker is making a batch of bran loaves and raisin loaves. One recipe requires $4\frac{7}{8}$ cups of flour and the other $3\frac{6}{8}$ cups of flour. How much flour is needed?

I think it's almost 9
$4\frac{7}{8} + 3\frac{6}{8}$
is the same as
$5 + 3\frac{5}{8}$
That's $8\frac{5}{8}$ cups.

Describe another strategy you could use.

Calculate the answer to each of these in your head.

1. $2\frac{6}{8} + 4\frac{5}{8}$
2. $5\frac{5}{8} + 2\frac{7}{8}$
3. $4\frac{13}{16} + 3\frac{15}{16}$
4. $4\frac{2}{4} + 3\frac{3}{4}$
5. $3\frac{4}{5} + 2\frac{3}{5}$
6. $6\frac{11}{16} + 1\frac{14}{16}$

Adding mixed numbers with related denominators

Introduction

In this activity, the students investigate strategies for solving number sentences involving mixed numbers with related denominators, such as $1^3/_4 + 2^1/_2$. One mental strategy is described on the transparency, however the students need to discuss other ways of calculating the examples that are shown.

Procedure

1. The strategy described on OHT 5.4 relies on the students' understanding of compatible fractions. In addition, two fractions are said to be compatible if they add to one whole. Write $^2/_8$ on the board. Ask the students to tell you another fraction that can be added to give one whole. Tell the students that two fractions that total one whole are said to be compatible fractions. Challenge the students to write as many compatible fractions as they can in a given time, such as one minute. It is also essential that the students have many experiences identifying equivalent fractions.

2. Show the top of OHT 5.4 and invite a volunteer to read the problem to the class. Ensure the remainder of the OHT is covered. Challenge the students to calculate the answer in their heads, then call upon volunteers to share and explain their thinking. After finding an equivalent fraction, some students may apply a 'rounding-and-adjusting' strategy. For example, $1^3/_4 + 2\,^1/_2$ is the same as $2 + 3 - ^3/_4$. Others may use a 'left-to-right' strategy, adding the ones first then the fractional parts. In this way, $1^3/_4 + 2^1/_2$ is the same as $3 + 1^1/_4$. Be sure to raise and discuss these strategies if they are not suggested by the students.

3. Show and discuss the strategy shown in the speech balloons. Make sure the students understand the idea of using a compatible fraction, then discuss the number sentences at the bottom of the OHT. Remember there is no one correct strategy. The students should be encouraged to use the strategy that best suits their way of thinking.

4. Allow time for the students to complete Activity 5.4 in their Mastery Booklet.

Adding fractions

Grace's fudge recipe needs $1\frac{3}{4}$ cups of raw sugar and $2\frac{1}{2}$ cups of white sugar. How much sugar is needed?

$1\frac{3}{4} + 2\frac{1}{2}$
This looks hard.
I'll try making a
compatible fraction!

$2\frac{1}{2}$ *is the same as*
$2\frac{1}{4} + \frac{1}{4}$
$1\frac{3}{4} + 2\frac{1}{4} + \frac{1}{4}$
That's $4\frac{1}{4}$ *cups!*

Calculate the sum of each of these in your head.

1. $3\frac{1}{4} + 4\frac{7}{8}$
2. $2\frac{1}{8} + 4\frac{15}{16}$
3. $4\frac{1}{4} + 5\frac{7}{8}$
4. $6\frac{1}{2} + 2\frac{5}{8}$
5. $2\frac{1}{4} + 7\frac{15}{16}$
6. $1\frac{5}{6} + 3\frac{1}{2}$

Review

Procedure

1. Read the sentence at the top of OHT 5.5 to the students. Ensure the remainder of the OHT is covered. Allow time for the students to calculate the answers. For each example, encourage them to share their thinking, then ask them to suggest other number sentences they could solve using the same strategy.

2. Show the students the second part of the OHT. Ask them to calculate the answer to Example E in their heads. Read the speech balloons to begin a discussion about the strategy they used. Some students may use a 'left-to-right' strategy, adding the ones first then the fractional parts. In this way, $2^4/_6 + 3^1/_2$ is the same as $5 + 1^1/_6$. Others may choose to express $^1/_2$ as $^1/_6 + ^1/_6 + ^1/_6$ then use a compatible pair to make a whole number. For example, $2^4/_6 + 3^1/_2$ is the same as $2^4/_6 + 3 + ^1/_6 + ^1/_6 + ^1/_6$, or $3 + 3^1/_6$. Afterwards, ask the students to share why they adopted their strategy for this particular example. Remember, flexible thinking should be praised and encouraged.

3. Show the students the remainder of the OHT. For each example, ask volunteers to explain how they would calculate the answer mentally. At this stage you may want to share the strategy you would use.

Review

Explain how you calculate the answer to each of these in your head.

a) $\frac{5}{8} + \frac{7}{8}$

b) $\frac{1}{2} + \frac{5}{8}$

c) $\frac{3}{4} + \frac{1}{2}$

d) $4\frac{3}{5} + 2\frac{4}{5}$

Calculate this number sentence in your head.

e) $2\frac{4}{6} + 3\frac{1}{2}$

Calculate the answer to each of these in your head.

1. $\frac{3}{8} + \frac{4}{8}$
2. $1\frac{2}{5} + 4\frac{4}{5}$
3. $\frac{1}{2} + \frac{1}{4}$
4. $3\frac{1}{4} + 4\frac{7}{8}$
5. $2\frac{3}{4} + 1\frac{2}{4}$
6. $5\frac{1}{2} + 2\frac{6}{8}$
7. $\frac{7}{8} + \frac{3}{4}$
8. $2\frac{1}{8} + 3\frac{14}{16}$
9. $3\frac{8}{16} + 4\frac{9}{16}$
10. $\frac{11}{16} + \frac{7}{16}$

Name: ____________________

Write the answers.

1. 235 + 46 = __________
2. 545 + 38 = __________
3. 215 + 27 = __________
4. 35 + 326 = __________
5. 64 + 128 = __________
6. 26 + 458 = __________
7. 327 + 82 = __________
8. 146 + 73 = __________
9. 36 + 345 = __________
10. 48 + 254 = __________

Act 1.2

Write the answers.

1. 325 + 246 = __________
2. 146 + 327 = __________
3. 635 + 218 = __________
4. 326 + 447 = __________
5. 318 + 265 = __________
6. 255 + 351 = __________
7. 463 + 154 = __________
8. 342 + 287 = __________
9. 486 + 142 = __________
10. 354 + 262 = __________

Write the answers.

1. 25.3 + 13.6 = __________
2. 34.5 + 24.3 = __________
3. 15.4 + 25.2 = __________
4. 36.7 + 22.5 = __________
5. 45.1 + 23.9 = __________
6. 31.3 + 24.8 = __________
7. 65.2 + 31.9 = __________
8. 42.8 + 12.6 = __________
9. 13.7 + 22.9 = __________
10. 63.4 + 22.8 = __________

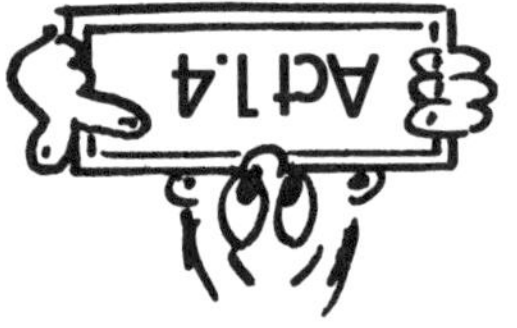

Write the answers.

1. $12.65 + $3.32 = __________
2. $6.54 + $2.38 = __________
3. $23.45 + $4.23 = __________
4. $36.12 + $3.47 = __________
5. $15.64 + $3.28 = __________
6. $7.22 + $12.36 = __________
7. $4.78 + $5.21 = __________
8. $45.35 + $3.52 = __________
9. $32.46 + $4.36 = __________
10. $53.48 + $5.25 = __________

Name: ____________

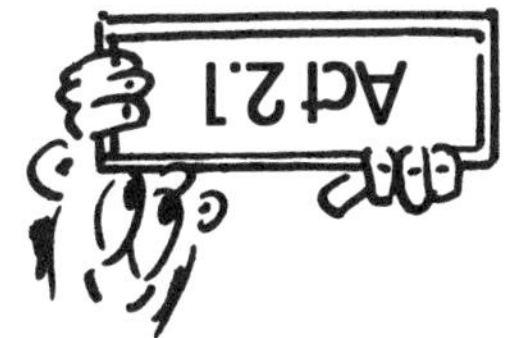

Write the answers.

1. **245 + 143 =** ______
2. **327 + 221 =** ______
3. **625 + 234 =** ______
4. **863 + 105 =** ______
5. **372 + 423 =** ______
6. **216 + 351 =** ______
7. **364 + 425 =** ______
8. **603 + 255 =** ______
9. **182 + 303 =** ______
10. **641 + 254 =** ______

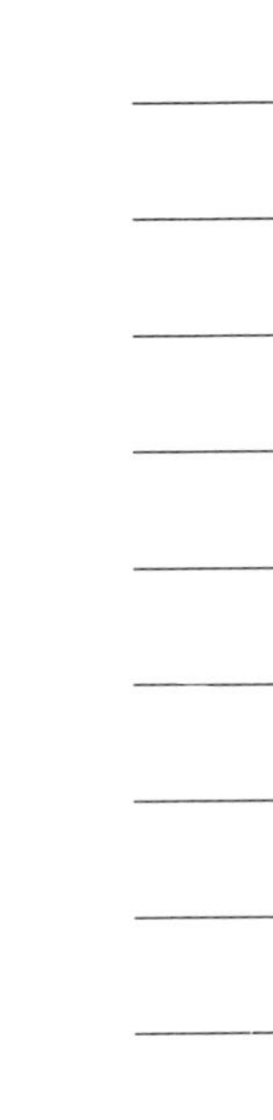

Write the answers.

1. **23.4 + 2.5 + 2.3 =** ______
2. **11.7 + 3.8 + 2.4 =** ______
3. **5.5 + 2.7 + 21.6 =** ______
4. **2.8 + 12.4 + 3.6 =** ______
5. **1.9 + 3.4 + 34.7 =** ______
6. **4.6 + 3.8 + 25.4 =** ______
7. **27.3 + 5.2 + 4.8 =** ______
8. **9.1 + 30.4 + 2.7 =** ______
9. **15.6 + 2.3 + 0.7 =** ______
10. **1.3 + 12.0 + 3.8 =** ______

Write the answers.

1. **327 + 256 =** ______
2. **436 + 248 =** ______
3. **605 + 276 =** ______
4. **624 + 359 =** ______
5. **253 + 463 =** ______
6. **762 + 155 =** ______
7. **483 + 246 =** ______
8. **456 + 387 =** ______
9. **395 + 256 =** ______
10. **573 + 288 =** ______

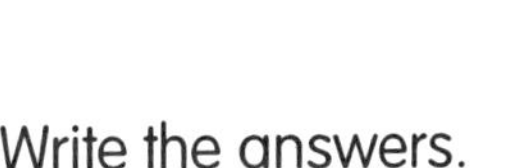

Write the answers.

1. **324 + 33 + 28 =** ______
2. **66 + 18 + 314 =** ______
3. **18 + 34 + 144 =** ______
4. **342 + 27 + 24 =** ______
5. **253 + 48 + 21 =** ______
6. **22 + 356 + 47 =** ______
7. **35 + 72 + 114 =** ______
8. **532 + 65 + 36 =** ______
9. **425 + 76 + 34 =** ______
10. **36 + 246 + 27 =** ______

Name: ____________

Write the answers.

1. $3.25 + $4.60 = ______
2. $1.60 + $5.35 = ______
3. $4.45 + $4.20 = ______
4. $7.10 + $2.55 = ______
5. $1.48 + $3.20 = ______
6. $5.67 + $2.30 = ______
7. $3.60 + $4.36 = ______
8. $6.52 + $2.40 = ______
9. $4.30 + $4.18 = ______
10. $1.60 + $3.33 = ______

Write the answers.

1. $3.60 + 85¢ = ______
2. $2.70 + 79¢ = ______
3. 48¢ + $5.80 = ______
4. $6.55 + 89¢ = ______
5. $7.95 + 48¢ = ______
6. 45¢ + $3.90 = ______
7. $7.68 + 49¢ = ______
8. $5.85 + 89¢ = ______
9. 65¢ + $3.78 = ______
10. 75¢ + $4.87 = ______

Write the answers.

1. $13.55 + $2.24 = ______
2. $23.35 + $4.62 = ______
3. $45.15 + $3.43 = ______
4. $5.42 + $13.37 = ______
5. $4.32 + $24.46 = ______
6. $2.61 + $42.33 = ______
7. $26.34 + $3.15 = ______
8. $3.64 + $22.35 = ______
9. $12.56 + $4.22 = ______
10. $35.12 + $1.45 = ______

Write the answers.

1. $22.45 + $3.37 = ______
2. $16.25 + $2.48 = ______
3. $32.65 + $4.26 = ______
4. $5.55 + $13.29 = ______
5. $4.34 + $33.47 = ______
6. $6.48 + $22.44 = ______
7. $24.65 + $3.77 = ______
8. $45.38 + $2.84 = ______
9. $4.56 + $23.75 = ______
10. $12.18 + $3.96 = ______

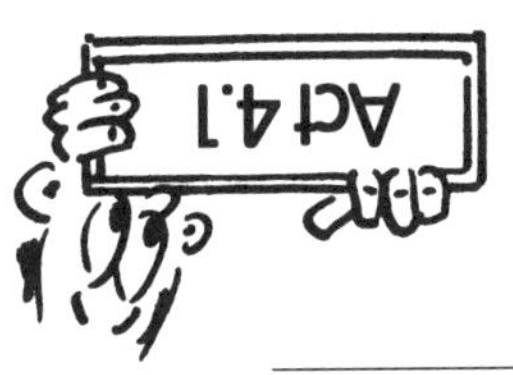

Name: ____________

Write the answers.

1. **248 + 398 =** ______
2. **197 + 147 =** ______
3. **396 + 249 =** ______
4. **452 + 349 =** ______
5. **295 + 447 =** ______
6. **354 + 198 =** ______
7. **253 + 148 =** ______
8. **297 + 349 =** ______
9. **154 + 247 =** ______
10. **345 + 297 =** ______

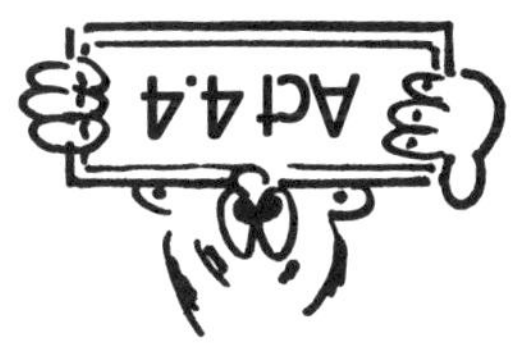

Write the answers.

1. **$12.48 + $16.45 =** ______
2. **$24.49 + $12.47 =** ______
3. **$34.46 + $21.48 =** ______
4. **$21.47 + $22.49 =** ______
5. **$30.45 + $32.45 =** ______
6. **$41.48 + $17.47 =** ______
7. **$15.49 + $12.49 =** ______
8. **$23.46 + $23.48 =** ______
9. **$51.48 + $11.48 =** ______
10. **$26.47 + $13.47 =** ______

Act 4.2

Write the answers.

1. **21.7 + 26.8 =** ______
2. **32.9 + 12.8 =** ______
3. **44.8 + 13.9 =** ______
4. **36.9 + 31.9 =** ______
5. **14.8 + 14.7 =** ______
6. **51.8 + 27.8 =** ______
7. **25.6 + 21.8 =** ______
8. **33.7 + 24.6 =** ______
9. **12.7 + 15.7 =** ______
10. **20.6 + 30.9 =** ______

Write the answers.

1. **$32.95 + $15.98 =** ______
2. **$13.98 + $12.95 =** ______
3. **$24.97 + $22.97 =** ______
4. **$16.96 + $31.99 =** ______
5. **$24.99 + $21.95 =** ______
6. **$16.97 + $11.99 =** ______
7. **$22.95 + $33.95 =** ______
8. **$20.98 + $14.98 =** ______
9. **$14.96 + $13.99 =** ______
10. **$35.99 + $42.99 =** ______

Name: ______________________

Write the answers.

1. $\frac{3}{5} + \frac{1}{5} =$ ______________

2. $\frac{3}{8} + \frac{4}{8} =$ ______________

3. $\frac{12}{16} + \frac{3}{16} =$ ______________

4. $\frac{3}{4} + \frac{2}{4} =$ ______________

5. $\frac{7}{8} + \frac{5}{8} =$ ______________

6. $\frac{13}{16} + \frac{7}{16} =$ ______________

Write the answers.

1. $2\frac{5}{8} + 6\frac{1}{2} =$ ______________

2. $3\frac{1}{2} + 1\frac{5}{6} =$ ______________

3. $4\frac{1}{4} + 5\frac{7}{8} =$ ______________

4. $4\frac{7}{8} + 3\frac{1}{4} =$ ______________

5. $2\frac{1}{4} + 7\frac{15}{16} =$ ______________

6. $4\frac{5}{16} + 2\frac{1}{8} =$ ______________

Write the answers.

1. $\frac{2}{4} + \frac{3}{8} =$ ______________

2. $\frac{1}{2} + \frac{1}{8} =$ ______________

3. $\frac{3}{8} + \frac{8}{16} =$ ______________

4. $\frac{3}{4} + \frac{3}{16} =$ ______________

5. $\frac{5}{8} + \frac{2}{4} =$ ______________

6. $\frac{10}{16} + \frac{4}{8} =$ ______________

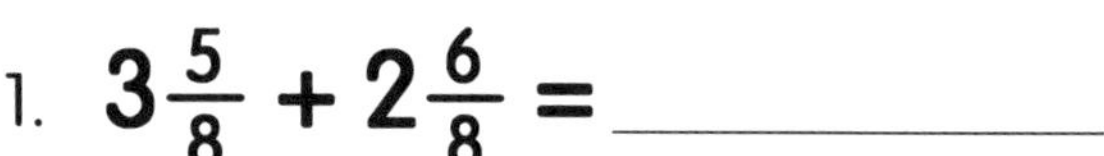

Write the answers.

1. $3\frac{5}{8} + 2\frac{6}{8} =$ ______________

2. $4\frac{3}{5} + 3\frac{4}{5} =$ ______________

3. $5\frac{13}{16} + 2\frac{15}{16} =$ ______________

4. $7\frac{3}{4} + 1\frac{3}{4} =$ ______________

5. $6\frac{7}{8} + 2\frac{6}{8} =$ ______________

6. $4\frac{7}{8} + 1\frac{7}{8} =$ ______________

Name:____

1. **There is a rule you can use to create equivalent fractions. Follow these steps to help you write equivalent fractions and identify the rule.**

- color a section to show the first fraction.
- draw lines between the points to reveal the second fraction.
- write what you did to the numerator and denominator to get the second fraction.

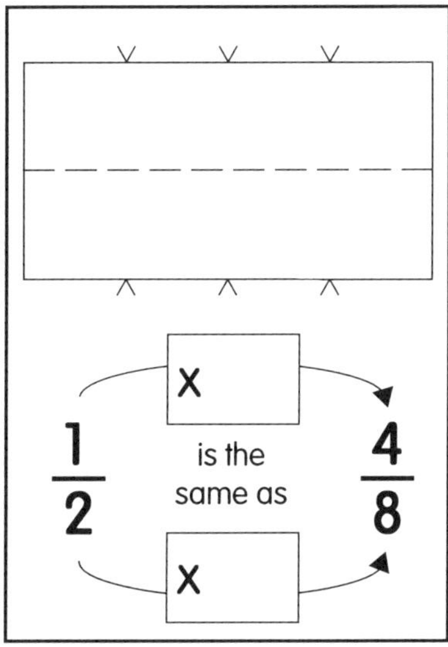

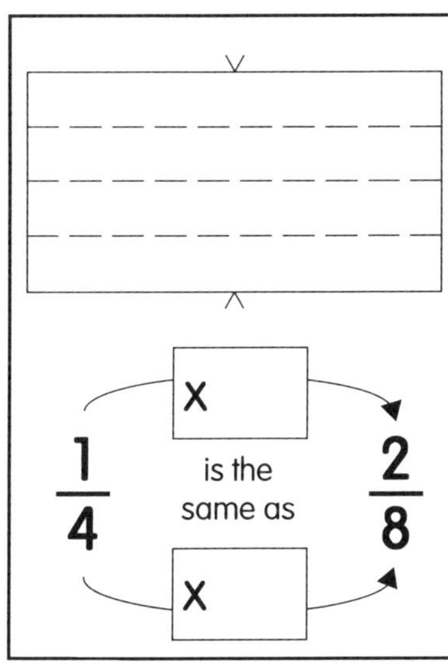

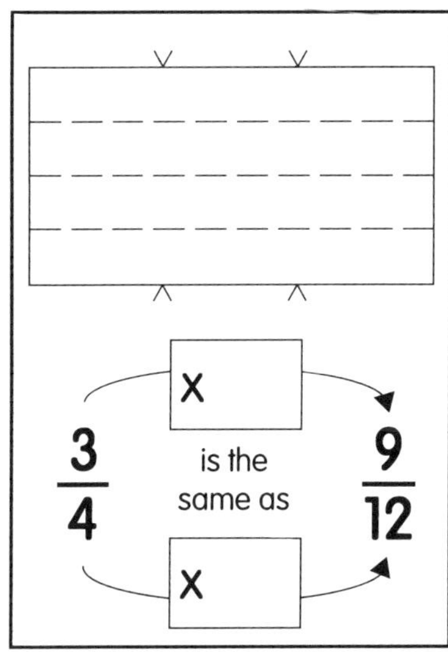

2. **Use the steps above to complete each of these.**

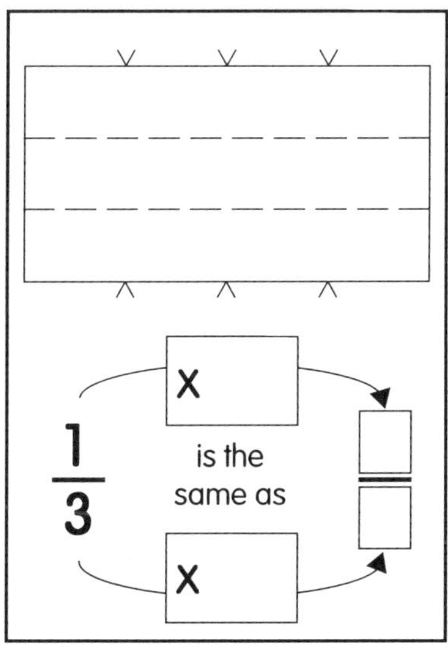

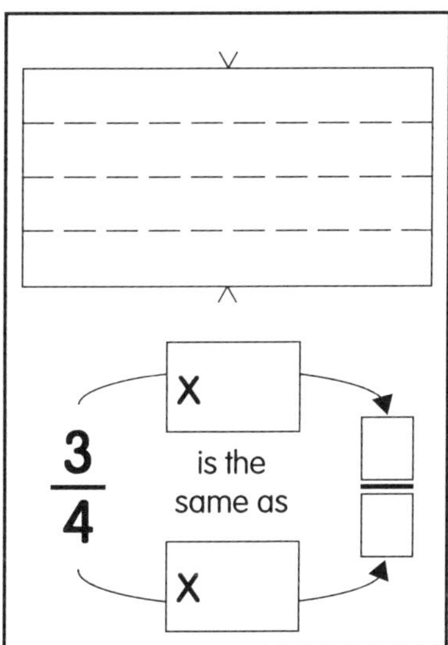

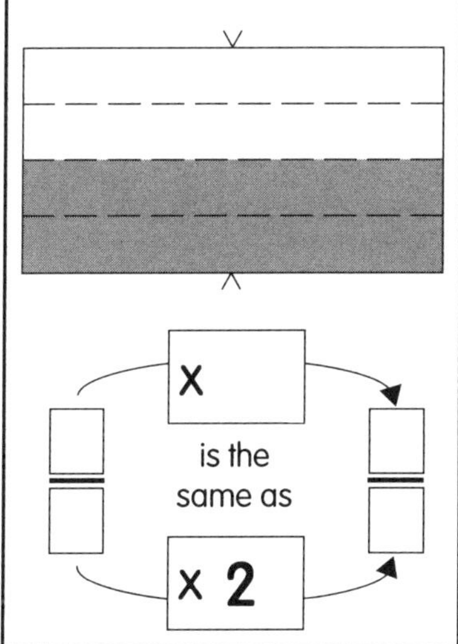

3. **Write a rule you could use to create equivalent fractions.**

____

____

Name: ______________________________

Calculate the answer to each of these.

1. $46 + 337 =$ ______
2. $64.3 + 23.8 =$ ______
3. $345 + 273 =$ ______
4. $\frac{3}{4} + \frac{1}{2} =$ ______
5. $54 + 38 + 343 =$ ______
6. $356 + 275 =$ ______
7. $1\frac{5}{6} + 3\frac{1}{2} =$ ______
8. $153 + 298 =$ ______
9. $172 + 116 =$ ______
10. 45¢ + \$45.70 = ______
11. \$5.74 + \$21.88 = ______
12. $\frac{5}{8} + \frac{2}{8} =$ ______
13. \$32.16 + \$3.62 = ______
14. $\frac{2}{5} + \frac{4}{5} =$ ______
15. \$1.10 + \$3.76 = ______
16. $3\frac{4}{5} + 2\frac{3}{5} =$ ______
17. \$65.56 + \$3.35 = ______
18. $35.9 + 23.9 =$ ______
19. $499 + 148 =$ ______
20. $4\frac{1}{4} + 5\frac{7}{8} =$ ______

---

21. \$15.66 + \$13.28 = ______
22. $33.4 + 5.5 + 6.2 =$ ______
23. \$42.98 + \$17.99 = ______
24. \$14.45 + \$24.45 = ______
25. \$32.97 + \$24.97 = ______

ANSWERS

Activity 1.1
OHT **1.** 461 **2.** 293 **3.** 182 **4.** 281 **5.** 292 **6.** 383 **7.** 607 **8.** 709 **9.** 418
Mastery Booklet **1.** 281 **2.** 583 **3.** 242 **4.** 361 **5.** 192 **6.** 484 **7.** 409 **8.** 219 **9.** 381 **10.** 302

Activity 1.2
OHT **1.** 382 **2.** 684 **3.** 563 **4.** 572 **5.** 772 **6.** 941 **7.** 618 **8.** 629 **9.** 619
Mastery Booklet **1.** 571 **2.** 473 **3.** 853 **4.** 773 **5.** 583 **6.** 606 **7.** 617 **8.** 629 **9.** 628 **10.** 616

Activity 1.3
OHT **1.** 39.8 **2.** 48.9 **3.** 58.7 **4.** 69.8 **5.** 47.9 **6.** 39.9 **7.** 59.3 **8.** 88.1 **9.** 76.2
Mastery Booklet **1.** 38.9 **2.** 58.8 **3.** 40.6 **4.** 59.2 **5.** 69 **6.** 56.1 **7.** 97.1 **8.** 55.4 **9.** 36.6 **10.** 86.2

Activity 1.4
OHT **1.** $15.79 **2.** $7.59 **3.** $27.77 **4.** $18.86 **5.** $36.78 **6.** $26.96 **7.** $18.94 **8.** $18.81 **9.** $68.91
Mastery Booklet **1.** $15.97 **2.** $8.92 **3.** $27.68 **4.** $39.59 **5.** $18.92 **6.** $19.58 **7.** $9.99 **8.** $48.87
9. $36.82 **10.** $58.73

Activity 1.5
OHT **a)** 464 **b)** 689 **c)** 59.8 **d)** 49.4 **e)** $28.95
1. 292 **2.** 39.7 **3.** 487 **4.** $15.87 **5.** 66.2 **6.** $36.88 **7.** 751 **8.** 39.1 **9.** $9.65 **10.** 281

Activity 2.1
OHT **1.** 359 **2.** 476 **3.** 497 **4.** 939 **5.** 288 **6.** 576 **7.** 857 **8.** 589 **9.** 526
Mastery Booklet **1.** 388 **2.** 548 **3.** 859 **4.** 968 **5.** 795 **6.** 567 **7.** 789 **8.** 858 **9.** 485 **10.** 895

Activity 2.2
OHT **1.** 792 **2.** 391 **3.** 571 **4.** 773 **5.** 626 **6.** 947 **7.** 631 **8.** 821 **9.** 631
Mastery Booklet **1.** 583 **2.** 684 **3.** 881 **4.** 983 **5.** 716 **6.** 917 **7.** 729 **8.** 843 **9.** 651 **10.** 861

Activity 2.3
OHT **1.** 295 **2.** 185 **3.** 399 **4.** 198 **5.** 299 **6.** 283 **7.** 523 **8.** 336 **9.** 435
Mastery Booklet **1.** 385 **2.** 398 **3.** 196 **4.** 393 **5.** 322 **6.** 425 **7.** 221 **8.** 633 **9.** 535 **10.** 309

Activity 2.4
OHT **1.** 29.1 **2.** 19.7 **3.** 28.3 **4.** 45.1 **5.** 36.3 **6.** 30
Mastery Booklet **1.** 28.2 **2.** 17.9 **3.** 29.8 **4.** 18.8 **5.** 40 **6.** 33.8 **7.** 37.3 **8.** 42.2 **9.** 18.6 **10.** 17.1

Activity 2.5
OHT **a)** 379 **b)** 631 **c)** 367 **d)** 235 **e)** 32.5
1. 569 **2.** 286 **3.** 395 **4.** 37.8 **5.** 188 **6.** 621 **7.** 19.6 **8.** 352 **9.** 43.8 **10.** 779

Activity 3.1
OHT **1.** $4.85 **2.** $8.95 **3.** $6.75 **4.** $7.77 **5.** $8.68 **6.** $6.76 **7.** $8.87 **8.** $7.68 **9.** $4.86
Mastery Booklet **1.** $7.85 **2.** $6.95 **3.** $8.65 **4.** $9.65 **5.** $4.68 **6.** $7.97 **7.** $7.96 **8.** $8.92 **9.** $8.48
10. $4.93

Activity 3.2
OHT **1.** $28.68 **2.** $38.79 **3.** $24.97 **4.** $18.78 **5.** $28.66 **6.** $38.48 **7.** $45.75 **8.** $35.87 **9.** $35.78
Mastery Booklet **1.** $15.79 **2.** $27.97 **3.** $48.58 **4.** $18.79 **5.** $28.78 **6.** $44.94 **7.** $29.49 **8.** $25.99
9. $16.78 **10.** $36.57

Activity 3.3
OHT **1.** $33.82 **2.** $25.94 **3.** $16.93 **4.** $26.91 **5.** $36.81 **6.** $15.71 **7.** $48.11 **8.** $28.33 **9.** $27.62
Mastery Booklet **1.** $25.82 **2.** $18.73 **3.** $36.91 **4.** $18.84 **5.** $37.81 **6.** $28.92 **7.** $28.42 **8.** $48.22
9. $28.31 **10.** $16.14

ANSWERS

Activity 3.4
OHT **1.** $9.79 **2.** $8.58 **3.** $9.46 **4.** $13.45 **5.** $46.15 **6.** $34.39 **7.** $17.24 **8.** $27.35 **9.** $37.11
Mastery Booklet **1.** $4.45 **2.** $3.49 **3.** $6.28 **4.** $7.44 **5.** $8.43 **6.** $4.35 **7.** $8.17 **8.** $6.74 **9.** $4.43 **10.** $5.62

Activity 3.5
OHT **a)** $7.87 **b)** $28.77 **c)** $18.62 **d)** $17.23 **e)** $24.38
1. $5.95 **2.** $10.01 **3.** $7.79 **4.** $8.25 **5.** $27.85 **6.** $7.77 **7.** $9.26 **8.** $18.62 **9.** $9.85 **10.** $17.24

Activity 4.1
OHT **1.** 694 **2.** 646 **3.** 647 **4.** 350 **5.** 601 **6.** 451 **7.** 902 **8.** 941 **9.** 602
Mastery Booklet **1.** 646 **2.** 344 **3.** 645 **4.** 801 **5.** 742 **6.** 552 **7.** 401 **8.** 646 **9.** 401 **10.** 642

Activity 4.2
OHT **1.** 54.5 **2.** 37.7 **3.** 49.3 **4.** 39.5 **5.** 38.5 **6.** 48.6 **7.** 66.4 **8.** 59.8 **9.** 39.7
Mastery Booklet **1.** 48.5 **2.** 45.7 **3.** 58.7 **4.** 68.8 **5.** 29.5 **6.** 79.6 **7.** 47.4 **8.** 58.3 **9.** 28.4 **10.** 51.5

Activity 4.3
OHT **1.** $39.90 **2.** $39.95 **3.** $57.94 **4.** $37.96 **5.** $44.94 **6.** $60.97
Mastery Booklet **1.** $48.93 **2.** $26.93 **3.** $47.94 **4.** $48.95 **5.** $46.94 **6.** $28.96 **7.** $56.90 **8.** $35.96 **9.** $28.95 **10.** $78.98

Activity 4.4
OHT **1.** $38.95 **2.** $55.94 **3.** $48.97 **4.** $47.92 **5.** $33.96 **6.** $38.90
Mastery Booklet **1.** $28.93 **2.** $36.96 **3.** $55.94 **4.** $43.96 **5.** $62.90 **6.** $58.95 **7.** $27.98 **8.** $46.94 **9.** $62.96 **10.** $39.94

Activity 4.5
OHT **a)** 645 **b)** 36.5 **c)** $38.93 **d)** $36.97 **e)** $47.90
1. 38.5 **2.** $37.96 **3.** $29.93 **4.** 646 **5.** $27.94 **6.** 445 **7.** $56.92 **8.** $43.95 **9.** 503 **10.** 49.7

Activity 5.1
OHT **1.** $\frac{4}{5}$ **2.** $\frac{7}{8}$ **3.** $\frac{15}{16}$ **4.** $\frac{13}{8}$ or $1\frac{5}{8}$ **5.** $\frac{5}{4}$ or $1\frac{1}{4}$ **6.** $\frac{7}{5}$ or $1\frac{2}{5}$ **7.** $\frac{18}{16}$ or $1\frac{2}{16}$ **8.** $\frac{10}{8}$ or $1\frac{2}{8}$ **9.** $\frac{6}{5}$ or $1\frac{1}{5}$
Mastery Booklet **1.** $\frac{4}{5}$ **2.** $\frac{7}{8}$ **3.** $\frac{15}{16}$ **4.** $\frac{5}{4}$ or $1\frac{1}{4}$ **5.** $\frac{12}{8}$ or $1\frac{4}{8}$ **6.** $\frac{20}{16}$ or $1\frac{4}{16}$

Activity 5.2
OHT **1.** $\frac{7}{8}$ **2.** $\frac{7}{8}$ **3.** $\frac{10}{16}$ **4.** $\frac{5}{4}$ or $1\frac{1}{4}$ **5.** $\frac{11}{8}$ or $1\frac{3}{8}$ **6.** $\frac{18}{16}$ or $1\frac{2}{16}$ **7.** $\frac{12}{8}$ or $1\frac{4}{8}$ **8.** $\frac{19}{16}$ or $1\frac{3}{16}$ **9.** $\frac{22}{16}$ or $1\frac{6}{16}$
Mastery Booklet **1.** $\frac{7}{8}$ **2.** $\frac{5}{8}$ **3.** $\frac{14}{16}$ **4.** $\frac{15}{16}$ **5.** $\frac{9}{8}$ or $1\frac{1}{8}$ **6.** $\frac{18}{16}$ or $1\frac{2}{16}$

Activity 5.3
OHT **1.** $7\frac{4}{8}$ **2.** $8\frac{4}{8}$ **3.** $8\frac{12}{16}$ **4.** $8\frac{1}{4}$ **5.** $6\frac{2}{5}$ **6.** $8\frac{9}{16}$
Mastery Booklet **1.** $6\frac{3}{8}$ **2.** $8\frac{2}{5}$ **3.** $8\frac{12}{16}$ **4.** $9\frac{2}{4}$ **5.** $9\frac{5}{8}$ **6.** $6\frac{6}{8}$

Activity 5.4
OHT **1.** $8\frac{1}{8}$ **2.** $7\frac{1}{16}$ **3.** $10\frac{1}{8}$ **4.** $6\frac{1}{8}$ **5.** $10\frac{3}{16}$ **6.** $5\frac{2}{6}$
Mastery Booklet **1.** $9\frac{1}{8}$ **2.** $5\frac{2}{6}$ **3.** $10\frac{1}{8}$ **4.** $8\frac{1}{8}$ **5.** $10\frac{3}{16}$ **6.** $6\frac{7}{16}$

Activity 5.5
OHT **a)** $\frac{12}{8}$ or $1\frac{4}{8}$ **b)** $\frac{9}{8}$ or $1\frac{1}{8}$ **c)** $\frac{5}{4}$ or $1\frac{1}{4}$ **d)** $7\frac{2}{5}$ **e)** $6\frac{1}{6}$
1. $\frac{7}{8}$ **2.** $6\frac{1}{5}$ **3.** $\frac{3}{4}$ **4.** $8\frac{1}{8}$ **5.** $4\frac{1}{4}$ **6.** $8\frac{2}{8}$ **7.** $\frac{13}{8}$ or $1\frac{5}{8}$ **8.** 6 **9.** $8\frac{1}{16}$ **10.** $\frac{18}{16}$ or $1\frac{2}{16}$

BLM 7
1. 383 **2.** 88.1 **3.** 618 **4.** $\frac{5}{4}$ or $1\frac{1}{4}$ **5.** 435 **6.** 631 **7.** $4\frac{8}{6}$ or $5\frac{2}{6}$ **8.** 451 **9.** 288 **10.** $46.15 **11.** $27.62
12. $\frac{7}{8}$ **13.** $35.78 **14.** $\frac{6}{5}$ or $1\frac{1}{5}$ **15.** $4.86 **16.** $5\frac{7}{5}$ or $6\frac{2}{5}$ **17.** $68.91 **18.** 69.8 **19.** 647 **20.** $9\frac{9}{8}$ or $10\frac{1}{8}$
21. $28.94 **22.** 45.1 **23.** $60.97 **24.** $38.90 **25.** $57.94